D1299556

SCIENCE AND TECHNOLOGY

RENNAY CRAATS

Weigl Publishers Inc.

Published by Weigl Publishers Inc.
350 5th Avenue, Suite 3304, PMB 6G
New York, NY 10118-0069
Website: www.weigl.com

Library of Congress Cataloging-in-Publication Data

Craats, Rennay.
Science and technology : USA past, present, future / Rennay Craats.
p. cm.
Includes index.
ISBN 978-1-59036-970-8 (hard cover : alk. paper) -- ISBN 978-1-59036-971-5 (soft cover : alk. paper)
1. Science--United States--History. 2. Technology--United States--History. I. Title.
Q127.U6C73 2009
506.73--dc22
2008023861

Printed in the United States of America
1 2 3 4 5 6 7 8 9 0 12 11 10 09 08

Weigl acknowledges Getty Images as its primary image supplier for this title.

EDITOR **Heather C. Hudak**
DESIGN **Terry Paulhus**

Science and Technology

Contents

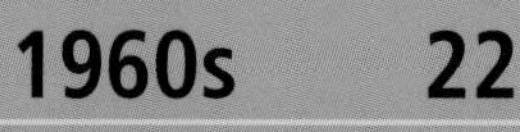

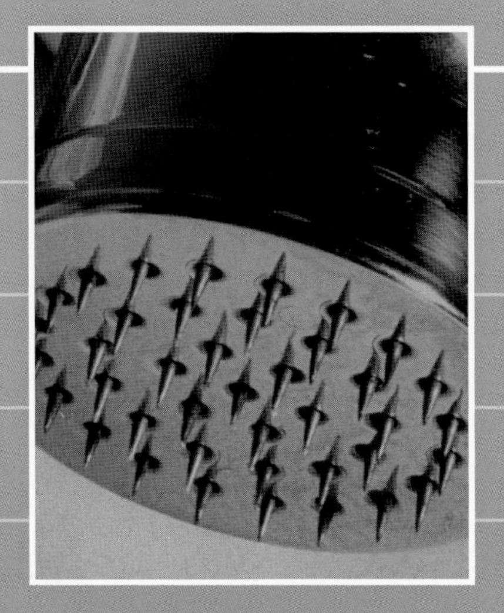

Science and Technology

Through The Years

Since the beginning of the 20th century, humans have made the most rapid scientific and technological progress in history. Inventors, scientists, and entrepreneurs led the rush to new advances. Many of these advances were made in the United States. From the invention of the automobile to the first steps on the Moon, scientific progress in the United States has shaped the world we live in today.

In the 20th century, Americans learned to fly, cure certain diseases, travel to space, and split the **atom**. Complex and powerful systems of communication were developed, linking the United States with every corner of the world.

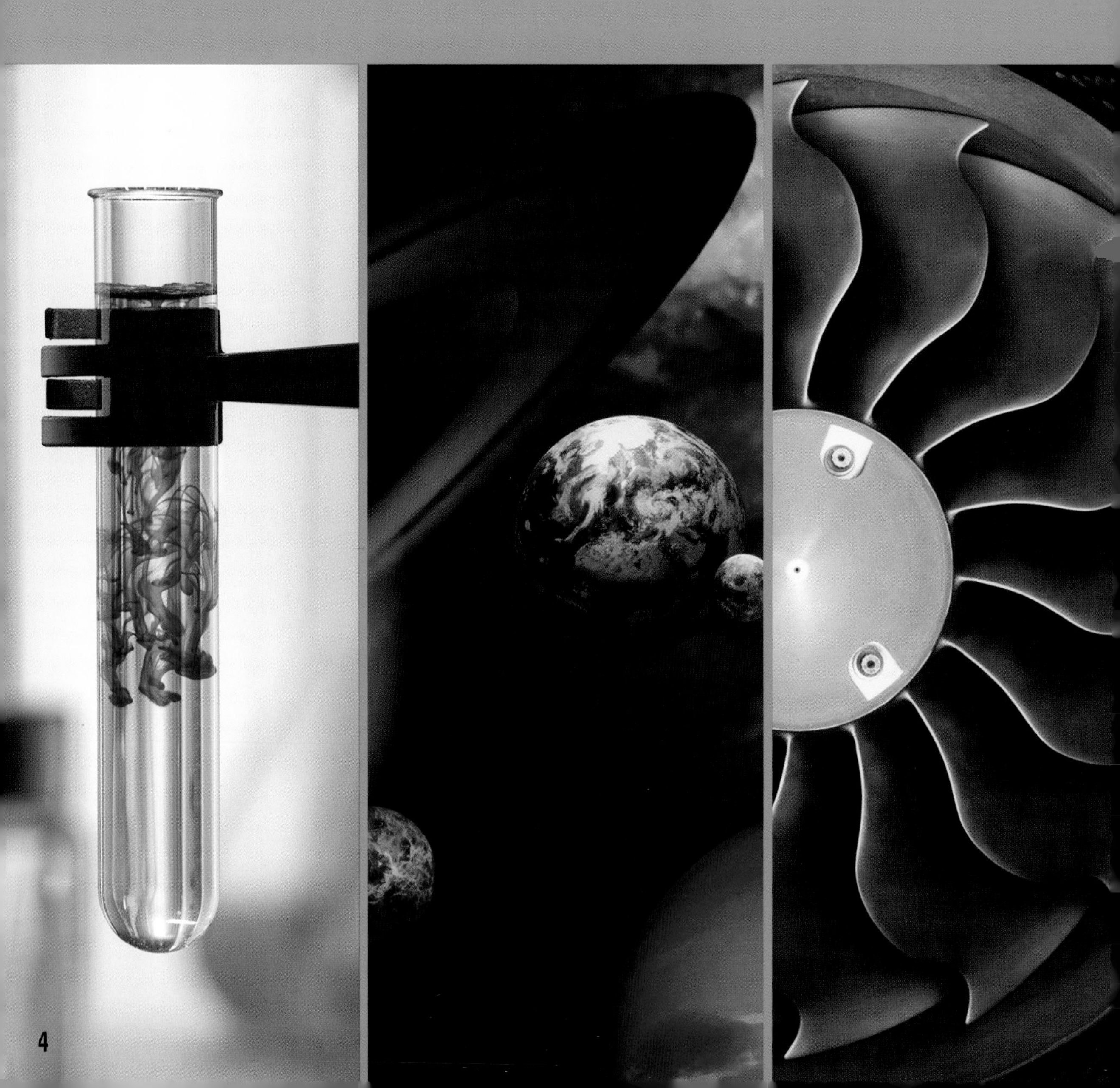

Americans went from viewing the world through a newspaper to sending messages via televisions and computer screens. Today, many U.S. citizens lead richer, longer, healthier lives because of the advances made in science and technology over the past 100 years.

As the United States moves into the 21st century, scientific and technological progress continues to make major strides in many areas. The hunger for knowledge continues to drive the best and brightest minds in the United States to build a brighter future this nation and of the world.

Science and Technology 2000s

2001

Artificial Heart

Robert Tools became the first person to receive a completely **self-contained** artificial heart. Doctors from the Jewish Hospital and University of Louisville, Kentucky, implanted the heart in Tools during a 7-hour procedure. Earlier attempts at developing a viable artificial heart had failed. Previous "hearts" used tubes and lines to connect bedridden patients to a console. They were used to keep patients alive while they waited for a heart transplant. The heart Tools received, called AbioCor, provides a more permanent solution. The "heart" moves blood throughout the body by mimicking a heartbeat. Weighing about 2 pounds and made from plastic and titanium, the heart's motor is powered by a rechargeable battery. The battery is charged from a special console that sends power through the skin. Lines and tubes are not needed to connect patients to the console, so they are able to live a more active lifestyle. Initial trials hoped to extend patient lives from 30 to 60 days. Tools survived 151 days without a human heart—much longer than projected. This small step was a major medical breakthrough. Following numerous clinical trials, in 2006, AbioCor received approval from the U.S. Food and Drug Administration. Each year, the artificial heart can be used on a limited number of patients meeting strict guidelines.

Artificial Heart

2001

Web 2.0

In 2001, a group of computer **programmers**, artists, and businesspeople began to think about the Internet in new ways. Early websites were seldom interactive, and programs often did not work well. The group began creating content specifically designed for the strengths of the Internet. They

Web 2.0

2001
NASA lands the first spacecraft on an asteroid.

2002
The first experiment on direct electrical communication between two human brains is done.

wanted to take advantage of the Internet's best features. Called "Web 2.0," this new approach resulted in extremely interactive websites and programs that allow users to create their own content and share it with other users. This new style of design paved the way for some of the most popular sites and programs on the Internet today, including YouTube, flickr, Facebook, and Wikipedia. These sites allow users to share videos, pictures, music, and writing without needing to know how to program their own website. The new innovations developed for Web 2.0 expanded the ability for people of different ages, cultures, and attitudes around the world to communicate and share ideas no matter their technical skill.

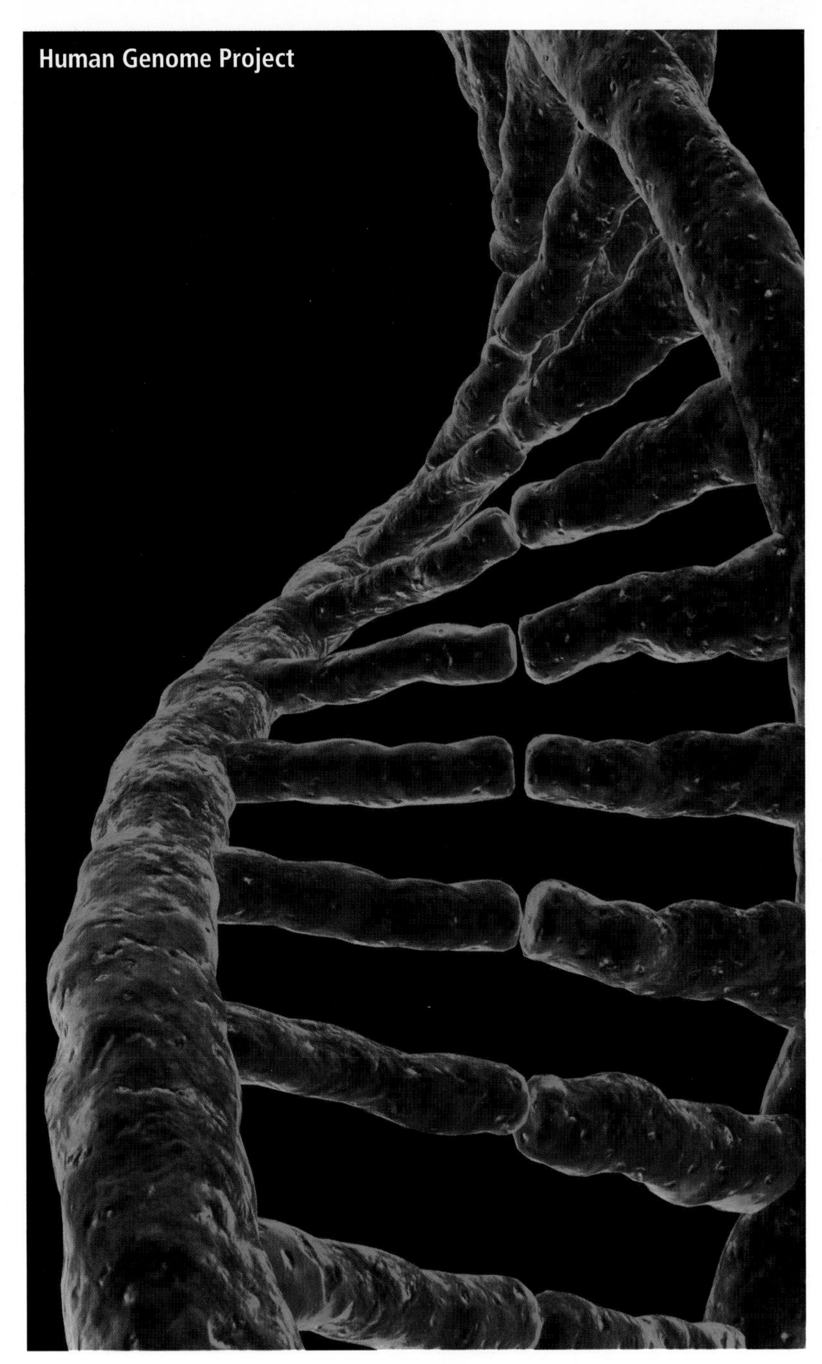
Human Genome Project

2003

Human Genome Project

After 13 years of research, the Human Genome Project concluded in 2003. The goals of the project included the identification of all of the **genes** in human DNA and finding the order of chemical base pairs. Funded by the United States Department of Energy and the U.S. National Institutes of Health, the project began in 1990 with the intent to complete research by 2005. However, advances in technology increased the speed of the project. In 2000, a rough draft of the human genome was released. Three years later, a finished version was made public. In 2006, final papers were released, while additional analysis continues to be released. Scientists plan to use information from the Human Genome Project to identify and cure diseases, make better medicines, and produce new **vaccines**.

2003
The age of the universe is estimated to be 13.7 billion years.

2004
Remains of a species closely related to humans are found in Indonesia.

2005
The first human face transplant is performed.

2006

Pluto: A Dwarf Planet

The number of planets in the solar system was reduced from nine to eight. Pluto had been reclassified as a dwarf planet. Scientists often questioned if Pluto should be considered a planet. Its small size, **elliptical** orbit, and extreme distance from the Sun made Pluto different from other planets. These doubts resurfaced in the 1990s with the discovery of more icy objects beyond Neptune. These objects, along with Pluto, are located in a distant ring called the Kuiper Belt. In 2005, American **astronomer** Mike Brown discovered an object in the Kuiper Belt that is slightly larger than Pluto. Some felt this object, now called Eris, should become the tenth planet. Others felt it called into question Pluto's classification as a planet. To resolve this debate, the International Astronomical Union (IAU) redefined the term "planet." This new definition placed both Pluto and Eris in the dwarf planet category, along with another object called Ceres.

2007

Airbus A380

Singapore Air took to the sky with the world's largest passenger airliner on October 25, 2007. The Airbus A380 can transport up to 525 passengers in comfort and style. Its double-deck cabin allows for wider seats and aisles than other commercial aircraft. As well, the A380 has 50 percent less cabin noise than other large aircraft, larger windows and cargo bins, and 2 feet of extra headroom. In addition to being larger than other aircraft, A380 has many features that benefit science and the environment. The aircraft burns 17 percent less fuel than the other large airliners, and it produces less than half of the target carbon dioxide emissions per passenger for European cars. To further reduce fuel consumption and gaseous emissions, the A380 was built using lightweight materials, aerodynamic innovations that decrease drag, and state-of-the-art engines.

2006
NASA launches the first space probe intended to observe Pluto.

2007
Scientists find evidence of water vapor on a planet outside the solar system.

Airbus A380

Into the Future

Scientists and engineers are always developing new technology. This may include advances in space observation, transportation, communications, and medicine. Think about discoveries made in the first years of the 21st century. What do you think scientists will discover next?

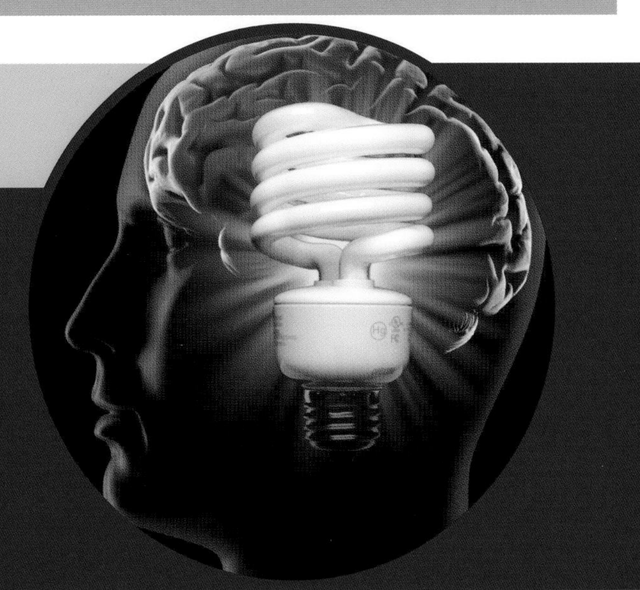

2008

The world's largest particle physics laboratory begins experiments.

2009

The United States is scheduled to begin broadcasting all television signals digitally.

Science and Technology

1990s

The Eyes of Hubble

1990

The Eyes of Hubble

In 1990, decades of planning paid off. The Hubble telescope was finally put in orbit 370 miles above Earth. This $1.5 billion technological eye was sent into space to explore distant parts of the universe. NASA had high hopes for the telescope, but it was also faced with high costs. The telescope's mirror had been ground to the wrong curvature, which meant that blurry images were sent back to Earth. Within three years, three of Hubble's **gyroscopes** had failed, and faulty solar panel supports trembled so badly in temperature changes that they could destroy the whole telescope at any moment. Despite these problems, the telescope managed to gather new information about the universe. For this reason, NASA decided to repair the mechanism. The shuttle *Discovery* blasted off in 1993, and crew members made the telescope as good as new. By then, the next generation of telescopes was beginning to appear. These new instruments gave more power and scope at a much lower cost.

Gene Therapy

1990

Gene Therapy

Scientists had studied many gene diseases, but few were as awful as ADA deficiencies. This rare condition was caused by a defect in the gene that tells cells to produce adenosine deaminase, an **enzyme** that stops the buildup of toxins that can

1991

1992

destroy the **immune system**. A drug was created to battle this disease in the eighties, but the expensive treatments did not always work. In September 1990, scientists made a breakthrough. A 4-year-old ADA sufferer became the first person to have her cells repaired or altered using one of the thousands of genes in DNA as a form of treatment. This successful gene therapy led to many experiments with other diseases, including cancer, AIDS, and cystic fibrosis.

1991

Biosphere Built

In October 1991, four men and four women took part in an incredible experiment. They started their two-year stay in a huge dome in the Arizona desert. A 3-acre sealed steel and glass enclosure was called Biosphere II. Inside Biosphere II were many miniature ecosystems, including a rain forest, a savanna, and an ocean. Also present were 3,800 species of plants and animals. The project was privately funded by Texas oil tycoon Edward Bass. He fronted the $150 million to get the project underway. Biosphere II was not as successful as everyone had hoped it would be. The crops within the enclosure failed, and in 1992, fresh air had to be pumped into the oxygen-starved dome. The eight residents abandoned the project in 1993.

Biosphere Built

1993
The term "spam" is first used to describe unwanted emails.

1994
One of the first Internet search engines, Yahoo!, is created.

1995
The Java programming language is released.

Internet Explosion

1993

Internet Explosion

The U.S. National Science Foundation was the main financier of the worldwide computer grid called the Internet. The Internet had begun as a military system in 1969 and was used to link researchers' computers. Then, it added libraries, universities, and other government departments. The military established 12,000 networks in forty-five countries. By 1993, 15 million people were using the Internet to research or chat with other "cybernauts." At this time, the foundation introduced a fast new system called T3. This system could handle 45 million bits of information per second. A revised T3 was soon operating at thirty times that speed. The communications capacity of this network grew enormously as the Internet spread throughout the world. More and more people became aware of "cyberspace" and the ability to use a personal computer to access information within seconds. With the Internet came instant messaging. Email became the preferred way to communicate with friends, family, and business associates, whether across the hall or across the world.

1995

Probing Jupiter

NASA began its Galileo Project in 1989. It was an unmanned mission to gather information about the planet Jupiter and its surroundings. With the launch of the space shuttle, *Atlantis*, the project was on its way. It had a probe set up to enter Jupiter's **atmosphere** and an orbiter that would continue to circle the planet. The probe photographed Jupiter, its moons, and radiation belts. On December 7, 1995, the Galileo probe reached Jupiter's atmosphere. For the first time, the planet's atmosphere was measured. The results of these tests allowed scientists to answer many questions that had been asked about the solar system's largest planet. The probe descended into Jupiter's atmosphere, and the orbiter continued to observe the planet and send data back to Earth.

1999

Lending a Hand

On January 24, 1999, a team of surgeons led by Dr. Jon Jones offered Matthew Scott a helping hand. They completed the first U.S. hand transplant. Scott had lost his left hand in a fireworks accident when he was 24 years old. Now, at the age of 38, he had a chance to make history

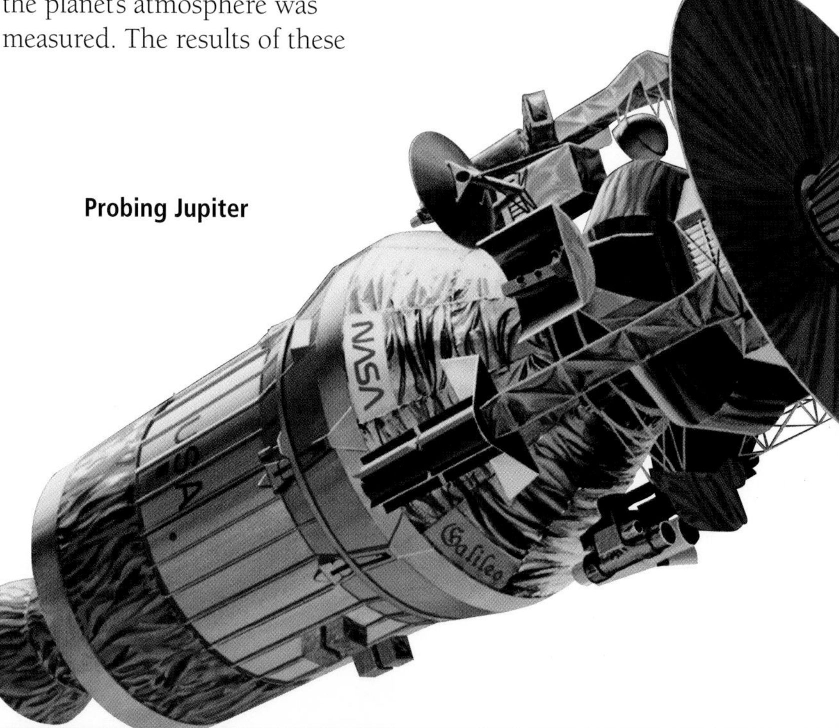

Probing Jupiter

1996 The first cloned animal, Dolly the sheep, is born.

1997 IBM's chess-playing computer defeats the world chess champion.

1998 John Glenn returns to space at 77 years of age.

and a new start. While surgeons had been reattaching severed limbs since the 1960s, they had never attached limbs from cadavers to living patients. This presented new obstacles and challenges. They had to prevent the patient's immune system from identifying the new limb as foreign and rejecting it. Using strong drugs, the surgeons were able to sidestep this problem. Within a year of the surgery, Scott was able to sense temperature, pressure, and pain with his new hand. He could also write, throw a baseball, and tie his shoelaces. The U.S. doctors were the second in the world to transplant a hand—the first hand transplant had occurred in France in 1998.

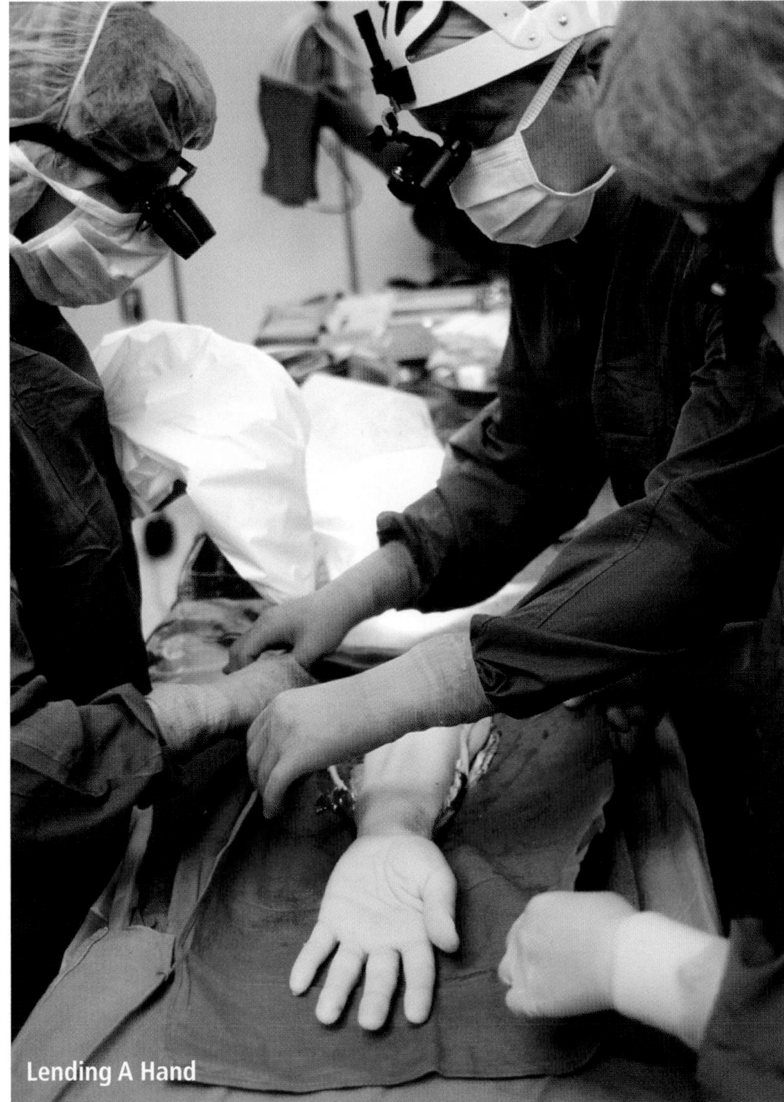
Lending A Hand

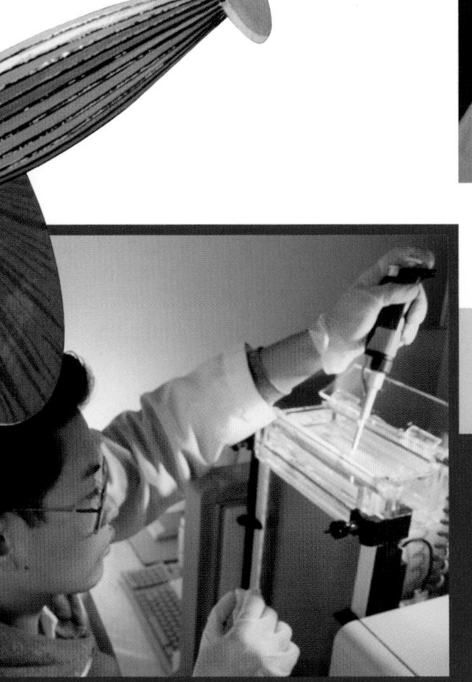

Into the Future

Genetic research is a powerful tool. It can help people discover why their bodies look and behave the way they do. It can be used to discover cures for diseases and heal ill people. Some people feel that such knowledge has the power to hurt people if it is misused. Think about the costs and the benefits of genetic research. How can it be used to help people?

1999

The largest bacteria on record is discovered. It is visible to the naked eye.

2000

Eric Kandel and his colleagues win the Nobel Prize for their work on the human nervous system.

Snapshots from Space

1980

Snapshots from Space

NASA launched two space probes to explore the solar system in 1977. In November 1980, the unmanned Voyager 1 probe sent incredible pictures back to Earth. In its journey through the solar system, the probe passed by Saturn. The photographs showed that the planet's rings were actually made of many smaller rings. It also discovered eight moons that were too tiny to be seen from Earth. Voyager 1 showed what scientists on Earth had seen only as spots of light. In 1989, Voyager 2 passed Neptune, sending back rare images of the planet. The space probes offered important information that astronomers would not have received otherwise.

1981

NASA launches the space shuttle *Columbia* for the first time.

1982

The first computer virus begins spreading across networks.

Computer Crazy

1981

Computer Crazy

Before the 1980s, most Americans could not afford a computer. In 1981, IBM introduced its personal computer (PC). These desktop computers were run by an Intel **microprocessor** that held the information to make the computer work. The computers used MS-DOS, which was Microsoft's operating system of programs used to run the machines. These components were soon reproduced and sold to other companies. IBM clones that ran just like authentic IBMs were quickly created and sold. IBM had not taken steps to prevent Intel and Microsoft from selling their products to companies that were looking to copy IBM's success. By 1984, about 3 million PCs were being sold each year. By the mid-1990s, about 90 percent of the world's computers were IBMs or copies.

1982

Handing Out Hearts

On December 2, 1982, Utah doctors gave Barney Clark an early Christmas gift—a new Jarvik-7 heart. A team of eighteen surgeons implanted the first artificial heart in a human at the University of Utah Medical Center in Salt Lake City. Before then, an artificial heart had been placed only in sheep and cattle. The operation took seven-and-a-half hours, and the plastic heart held up well. But complications set in soon after the operation. Clark suffered from seizures in the first week and had problems with his lungs and liver. This was likely because of less blood flow in these areas during surgery. The patient lived for 112 days after the operation. Over the next ten years, the artificial heart was improved upon and used for patients waiting for heart transplants.

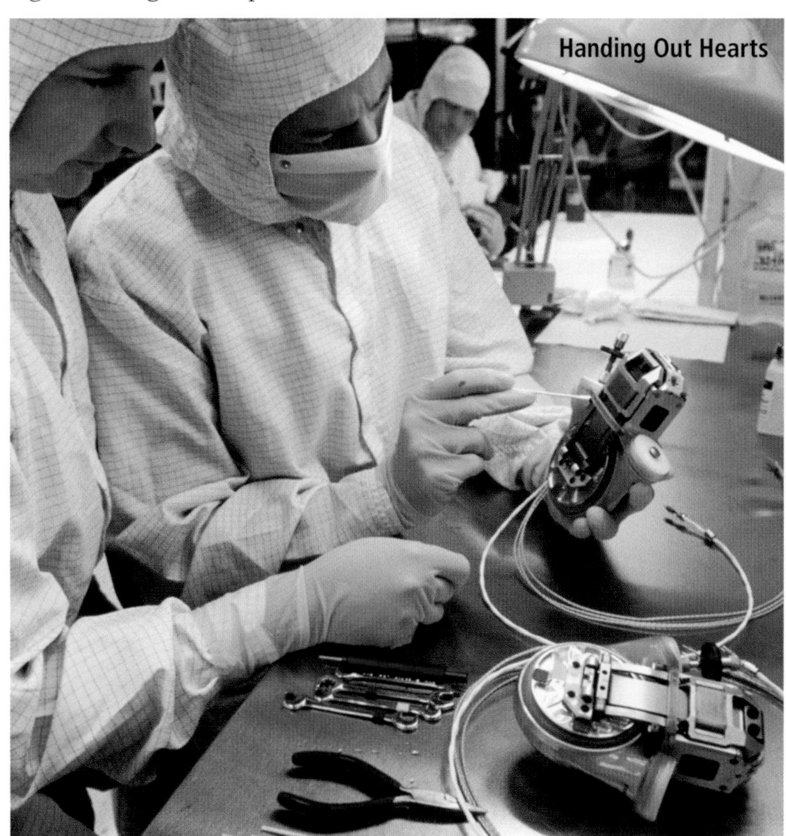
Handing Out Hearts

1983
A magazine releases instructions to create the first homemade computer.

1984
Astronauts make the first space walk without being tied down.

1985
The wreck of the RMS *Titanic* is located.

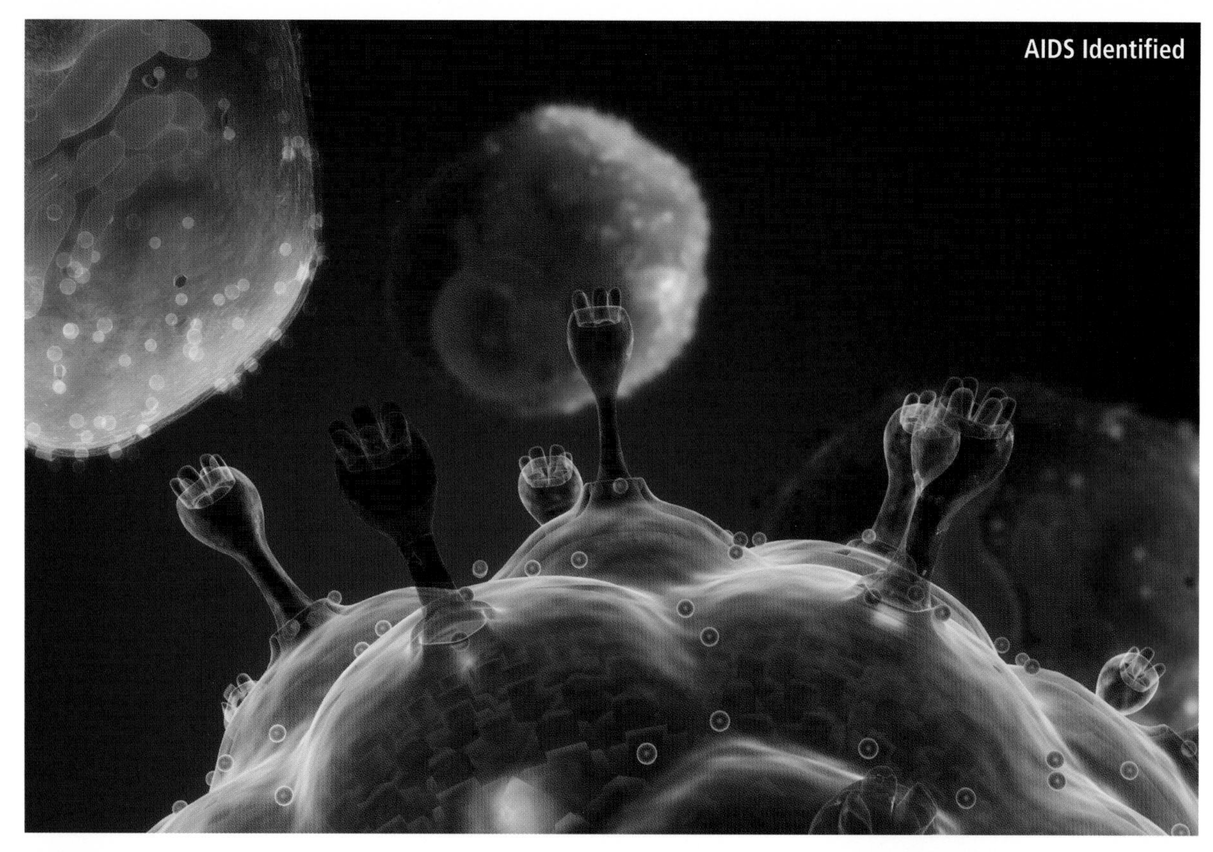

1982

AIDS Identified

In the early 1980s, mysterious illnesses were being reported around the world. Doctors could not explain why so many people were dying from treatable illnesses such as pneumonia and skin cancer.

By 1982, doctors had examined sufferers and realized that the problem lay with their immune systems. The immune system helps the body fight infections and disease. The new condition, called acquired immunodeficiency syndrome, or AIDS, prevented the immune system from doing its job. It was thought to be caused by HIV, or human immunodeficiency virus, in the body. There was a chance that people with HIV would develop AIDS.

By December 1982, 1,600 cases of AIDS had been reported worldwide. The number of cases was doubling every six months. HIV is passed through body fluids. Any activity that involves sharing body fluids is considered risky. People became terrified that they would develop AIDS. Many people, including Americans, started to alter their behavior to keep themselves safe.

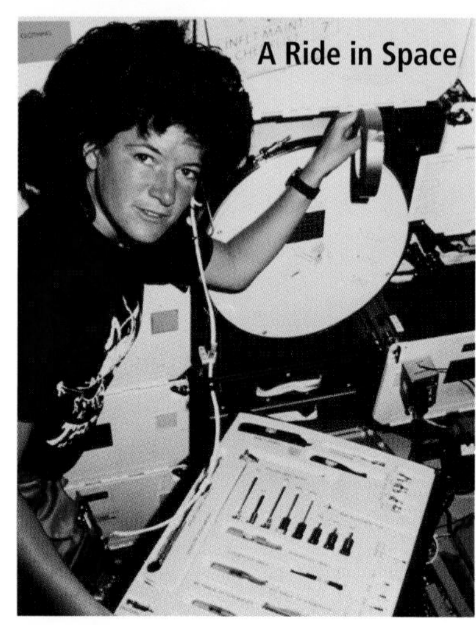

1983

A Ride in Space

Sally Ride was a professional tennis player before she became

1986
IBM creates the first laptop computer.

1987
A heart and lung transplant is performed for the first time.

1988
The first shuttle is sent into space after the *Challenger* explosion.

star-struck. She left sports to finish her Ph.D. in astrophysics.

On June 18, 1983, Ride became the first U.S. woman in space. During the six-day flight, Ride launched communications satellites and tested the shuttle's remote arm. She went into orbit again on October 5, 1984. A few years later, she was asked to join the commission investigating the *Challenger* explosion of 1986. Ride resigned from the space program in 1987. She became director of the California Space Institute at the Scripps Institution of Oceanography, as well as a physics professor at the University of California in San Diego. She was inducted into the National Women's Hall of Fame in 1988.

Virtual Invention

1984

Virtual Invention

In 1984, "virtual reality" became a reality. A twenty-four-year-old inventor named Jaron Lanier and his company VPL Research created the first virtual reality environment. He created the headset, gloves, and suits, along with the software for "cybernauts" to explore while playing the game. Participants could see, hear, and interact with a digital world. Early VR programs had their bugs, including a cartoonish world and some headaches and nausea from the headset. Yet people were still drawn to the possibilities that virtual reality presented. By the 1990s, many other companies were working on virtual reality. The military and airlines used the simulator concept to train pilots. VR games and other recreational three-dimensional movies were offered as well. Virtual reality took the computer experience to the next level.

Into the Future

Computers play a key role in many of the activities people do today. Communication, work, and entertainment are all affected by computers. Think about the ways you use computers in your daily life. Do you think they will be used more or less often in the future? What other ways might they be used?

1989

Cold fusion is observed by Martin Fleischmann and Stanley Pons at the University of Utah.

1990

The Hubble Telescope makes its way into orbit.

Science and Technology

1970s

Unlucky Thirteen

Being Green

Unlucky Thirteen

It was supposed to be simple—the third landing on the Moon. In April 1970, the spacecraft Apollo 13 drew millions of Americans to their televisions to see if the astronauts would make it home.

As Apollo 13 was nearing the Moon, one of the oxygen tanks had blown up and disabled the other. Astronauts relied on these tanks for breathing and to fuel the electrical systems. Without these reserves, the astronauts would likely not have enough oxygen to return to Earth. The crew transferred into the lunar module, Aquarius, which was meant just for landing on the Moon. It was not equipped to support three people. The crew then shut down their main command module to save energy until they needed to re-enter Earth's atmosphere. The lunar module had its own power and oxygen, and the astronauts used as little power as possible during their three-day return to Earth. They overcame near-freezing temperatures and too much carbon dioxide in the lunar module, and they landed safely in the Pacific Ocean on April 17. The entire country breathed a sigh of relief as the crew emerged from the module.

1970

Being Green

Concern about the health of the environment led to the first officially recognized Earth Day on April 22, 1970. This first Earth awareness day included seminars, parades, and recycling fairs. The federal government also realized the importance of protecting the environment.

In 1972, the Environmental Protection Agency was established. The United Nations assisted in raising awareness by holding an environmental conference in Sweden. This was not enough for some activists. Many people joined environmental groups and demanded a change in policies from companies that were polluting the environment.

1971

Intel releases the world's first microprocessor.

1972

President Richard Nixon orders the creation of the space shuttle program.

Green fever hit Americans hard, and more people insisted that resources be used responsibly. Activists argued that humans had a duty to manage the natural environment responsibly. Through their efforts, many international environmental policies have been developed.

Towering Above

1973

Towering Above

In 1973, Americans were at the top. The Sears Tower in Chicago was completed, becoming the tallest building in the world. The 110-story building was 1,454 feet high and spread across two city blocks. It held 4.5 million square feet of office space and used about as much power as a town of 35,000 people. The amazing Sears Tower was designed by the architectural firm Skidmore, Owings, and Merrill. It was built by Sears, Roebuck, and Company in less than two-and-a-half years. On a clear day, four states—Illinois, Indiana, Michigan, and Wisconsin—are visible from the observation deck on the 103rd floor. Before this structure was built, the Empire State Building in New York had held the title of tallest building.

1973
The United States launches its first space station, known as Skylab.

1974
The world's longest immersed subway tunnel opens in San Francisco.

1975
Atari makes the first home video game, *Pong*.

An Apple A Day
Steve Jobs

1976

An Apple a Day

Stephen Wozniak and Steven Jobs designed a computer circuit board in Jobs' California garage in 1976. The Apple I was sold without a monitor, keyboard, or casing. The following year, the company introduced the personal computer, Apple II, and it released another model, Apple III, in 1980. The Apple III did not sell well because it was very expensive and had hardware problems. Apple worked out the bugs in the system, and by 1982, it became the first personal-computer company to achieve annual sales of $1 billion. Apple continues to develop and market new personal computers.

Ingenious Discovery
Har Gobind Khorana

1976

Ingenious Discovery

Led by Dr. Har Gobind Khorana, a team of scientists at the Massachusetts Institute of Technology made history in 1976. They created the first human-made gene that could work inside a cell. Unlike previous genes, this one was completely manufactured—a natural gene was not used as a model. The biochemists had put together a gene that corrects mutations. This discovery was met with mixed feelings. Such a discovery could, down the road, help prevent hereditary diseases. It could also be used in genetic engineering. This made some Americans very nervous. The debate over genetic engineering carried through into the 21st century.

1976
The U.S. Air Force begins test flights of the F-16 Fighting Falcon.

1977
Scientists identify the cause Legionnaire's disease.

1978
The first people cross the Atlantic Ocean by balloon.

1977

Supersonic

In 1977, the world's first supersonic passenger plane took off from Kennedy Airport in New York. It held 100 passengers and flew twice as fast as other airplanes. After nearly ten years of technical glitches and financial problems, the *Concorde* was at last in the air. It had been a long, hard-won battle for the French and British builders. The *Concorde* was surrounded by controversy. Many people warned of environmental damage to Earth's atmosphere if such planes were permitted to fly. Others fought against the deafening sound the *Concorde* produced. Despite these concerns, the *Concorde* passed all requirements and began offering service across the Atlantic Ocean.

Into the Future

The concern for the environment people displayed on the first Earth Day continues to grow today. Many scientists and inventors are finding ways to live comfortable lives without harming the world. Think about ways that you can look after the environment in your everyday life.

1979

Television ownership worldwide grows to more than 300 million households.

1980

Proof is found that there has been life on Earth for about 3.5 billion years.

Science and Technology

1960s

1961

Keyboard Music

Milton Babbitt was a mathematician who turned his knowledge of numbers and formulas to music. He was one of the first people to use computer technology to study the structure of music. In 1959, Babbitt helped establish the Columbia-Princeton Electronic Music Center. His "Composition for Synthesizer" (1961) was one of the first pieces of electronic music. Other works include "Philomel" (1964) for soprano and magnetic tape, and "Concerti for Violin, Small Orchestra, and Synthesized Tape" (1976). Babbitt decided to use certain pitches, harmonies, and rhythms beforehand, and created what he called "total serialization" in his music. He was honored with a Pulitzer Special Citation for his work in 1982.

1962

Television in Space

In 1962, AT&T sent the first communications satellite into orbit. Telstar floated 500 to 3,500 miles above Earth. It received faint television signals, magnified them 10 billion times, and then sent them back to Earth. This allowed television audiences in the U.S. to receive programs from Europe, and Europeans to receive U.S. programming. The implications of what Telstar could do caused great debate over future satellites and ground stations—would the government develop high-orbit satellites and own the skies, or would AT&T be allowed to continue with its low-orbit satellites? To come out ahead of the Soviets in telecommunications, President Kennedy decided to back AT&T, making the company more successful than ever before.

1962

Space Heroes

On February 20, 1962, an American entered space aboard the spacecraft *Friendship*—John Glenn was the first American to circle the Earth. He orbited three times in less than five hours as Americans sat glued to their television sets watching him, but it was far from a perfect mission. Instead of firing small jets to keep the spacecraft in position,

Keyboard Music

1961

A chimpanzee is sent into space to test a capsule meant for human use.

1962

The Ranger 4 space probe lands on the Moon.

1963

The word "clone" is first used to describe the process of copying DNA sequences.

Friendship was firing large jets—and using up fuel that was needed to get home.

Then, ground station readings suggested that the craft's heat shield had been detached. This meant that *Friendship* and Glenn would burn up upon re-entering Earth's atmosphere. To overcome these problems, Glenn switched to manual-control systems to stop the big jets from firing. The worries about the heat shield proved false, and he landed safely in the Atlantic Ocean.

In June 1965, astronaut Edward White became the first American to walk in space. For twenty-two minutes, White floated in space attached by a cord to the Gemini 4 capsule. He used a gun that fired bursts of compressed oxygen to propel himself in the weightlessness of space. The flight commander, James McDivitt, took pictures of the historic moment. The astronauts returned to Earth as heroes.

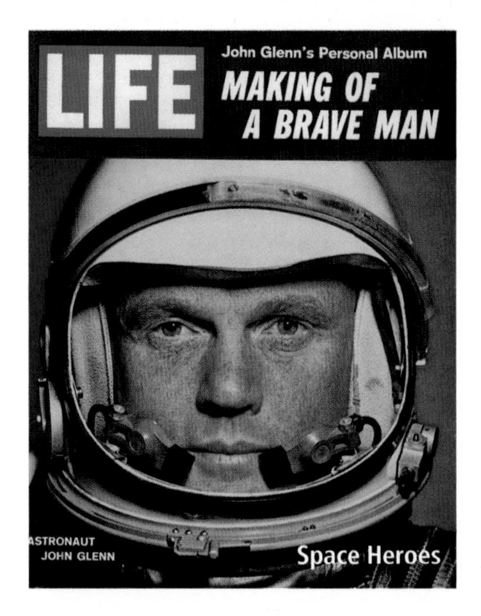

Space Heroes

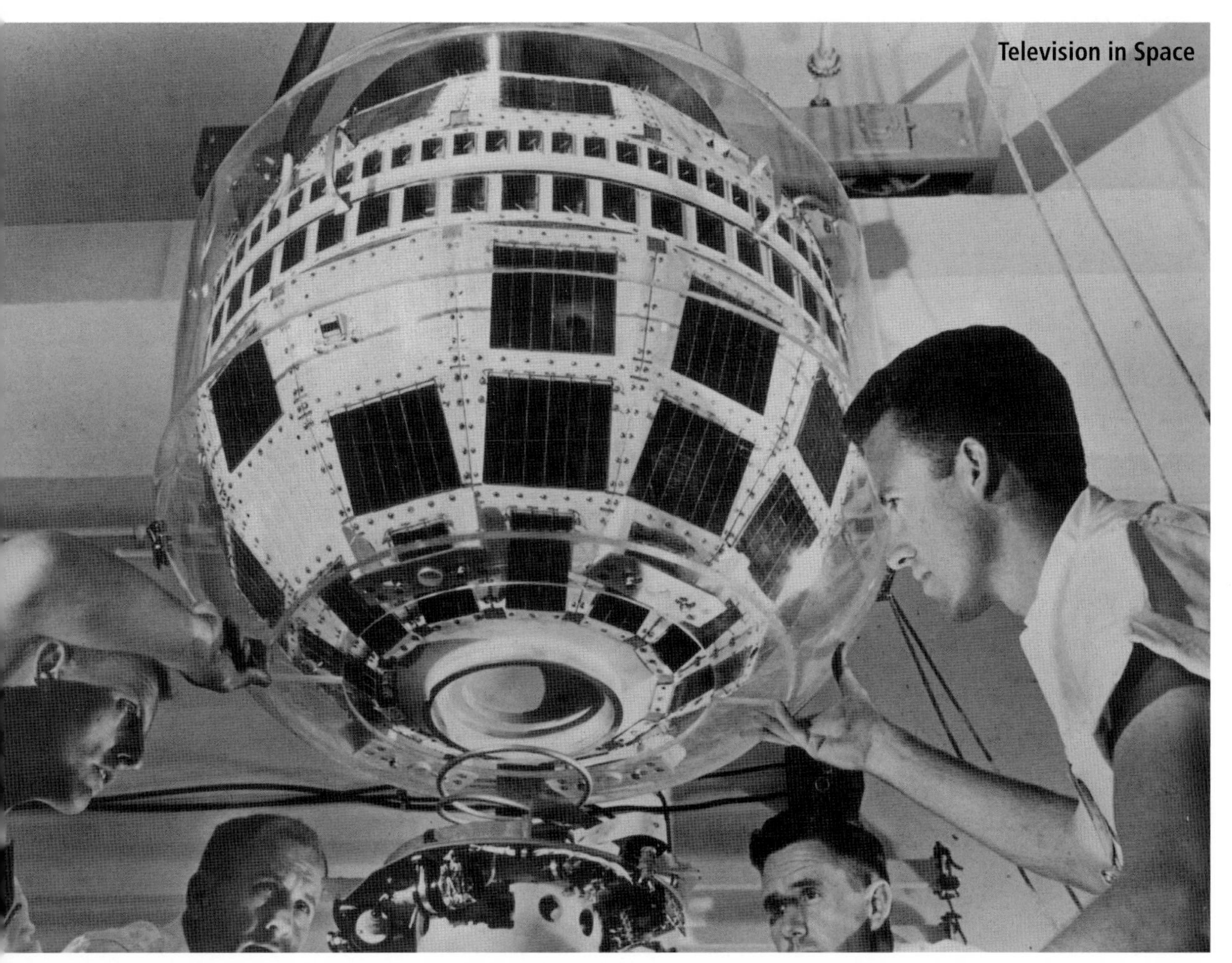
Television in Space

1964
The U.S. surgeon general announces that smoking may be harmful.

1965
Live television images are sent from the Ranger 9 space probe as it nears the Moon.

1969

The Eagle Has Landed

About 600 million people watched on television as astronauts broadcast their transmissions from the Moon. As John Kennedy had promised, Americans visited the Moon before the end of the decade. On July 16, 1969, Michael Collins, Neil Armstrong, and Edwin "Buzz" Aldrin, Jr., boarded Apollo 11 and headed for the Moon. Four days later, Aldrin and Armstrong climbed into the landing craft called the Eagle to complete their trip to the Moon. The astronauts were surprised by the huge boulders on the Moon's surface but successfully touched down. Armstrong stepped out first onto the powdery surface of the Moon, kangaroo-hopping around in the low-gravity atmosphere. Aldrin joined him nineteen minutes later. For more than two hours, they collected samples, set up instruments that would monitor the atmosphere, and planted a U.S. flag to show they were the first to explore the Moon. Armstrong and Aldrin reluctantly returned to their spacecraft and headed back to Earth. They safely splashed into the Pacific Ocean off the coast of Hawai'i on July 24 and took their place in the history books.

The Eagle Has Landed

1966
Construction of the Gateway Arch in St. Louis, Missouri is completed.

1967
A new type of rapidly-flashing star, a pulsar, is discovered.

1968
The first successful heart transplant is performed.

1962

Thalidomide Pulled From Market

Around the world, thalidomide was called "the sleeping pill of the century." What people did not know was how dangerous it was to unborn children. Pregnant women who took the drug put their unborn children at risk. Thalidomide caused up to 12,000 babies to be born disabled throughout the world. Many of these children died soon after birth. Other babies were born with flipper-like limbs instead of arms and legs. Often, their ears, eyes, and internal organs were also damaged. The drug was pulled from the market in 1962, but the harm had been done.

In the U.S., only about a dozen children were affected, thanks to Frances Oldham Kelsey, an FDA investigator. She turned down the drug after reading that it caused nerve inflammation and did not always aid in sleeping. She faced pressure to accept the drug for more than a year, but she refused. Even though the drug was not approved, some U.S. doctors got free samples from the makers. About 20,000 female patients in the U.S. used the drug. Once the world realized the disastrous effects of the drug, Kelsey was awarded a medal by President Kennedy.

Into the Future

People can see the effects of the space program every day in the news and information they receive from around the world. Global communication is made possible by satellites. What do you think the world would be like today if these satellites had not been invented? How would people communicate over distances?

1969

A computerized system is developed for aircraft that can replace a human navigator.

1970

The first remote-controlled robot explorer lands on the Moon.

Science and Technology 1950s

Nuclear Energy

1951

Nuclear Energy

Six years after atom bombs destroyed Hiroshima and Nagasaki, U.S. scientists found another way to use atomic power. On December 29, 1951, the U.S. Atomic Energy Commission announced that it had found a peaceful use for **nuclear energy**. Experiments at the Arco, Idaho, nuclear reactor had led to new sources of energy. The heat from the reactor boiled water, and the steam that resulted powered a turbine. Scientists had generated a steady stream of electricity using nuclear power. This success gave them hope that they had discovered an inexpensive energy source—one that would break the country's dependence on fossil fuels such as coal and petroleum. Nuclear power stations were later built in other countries, including France. However, the high costs of building and maintaining these plants prevented nuclear energy from becoming a main power source in many countries.

1952

Medical Hero

Thousands of children contracted poliomyelitis, or polio, each year in the 1940s and early 1950s. This disease attacked muscles and caused paralysis. There was no way to protect people against this

1951 The first UNIVAC I computer is given to the United States Census Bureau.

1952 Archaeologists discover the remains of a Viking ship near Boston.

1953 The structure of human DNA is first described.

contagious disease. Dr. Jonas Salk began experimenting with ways to prevent polio in 1947. Most vaccines were made from weakened but live strains of whatever disease they were fighting. This was too risky with polio. Salk used dead polio viruses to try to beat the disease. By 1952, the doctor had created a vaccine that stopped polio. He and his staff were injected with the vaccine, and public trials began in 1954. The following year, Salk's vaccine was licensed. It calmed worldwide fears of the disease. Americans watched with relief as the number of polio cases continued to drop.

1952

Power Unleashed

On November 1, 1952, a three-mile-wide cloud developed above the Marshall Islands in the northcentral Pacific Ocean. It was the first test of a hydrogen bomb. The bomb exploded with the power of 10 million tons of dynamite, and it was 500 times more powerful than the bomb dropped on Hiroshima. Physicist Edward Teller, leader of the project, was thrilled with the bomb's success, but not everyone shared his enthusiasm. Atomic scientists J. Robert Oppenheimer, who had directed the development of the first atomic bomb, and Enrico Fermi, who first thought of creating such a **thermonuclear** superbomb, were against the bomb on moral grounds. Oppenheimer was considered a threat and, after he was accused of being a communist, he lost his security clearance for the project. Despite concerns, President Truman ordered the Atomic Energy Commission to create the hydrogen bomb, or H-bomb, as quickly as possible. Americans tested three more H-bombs in 1954 and another seventeen in 1955.

Medical Hero Jonas Salk

Power Unleashed

1954

The first large-scale polio vaccinations take place in Pittsburgh.

1955

George Smith becomes the first person to survive ejecting from a jet traveling faster than the speed of sound.

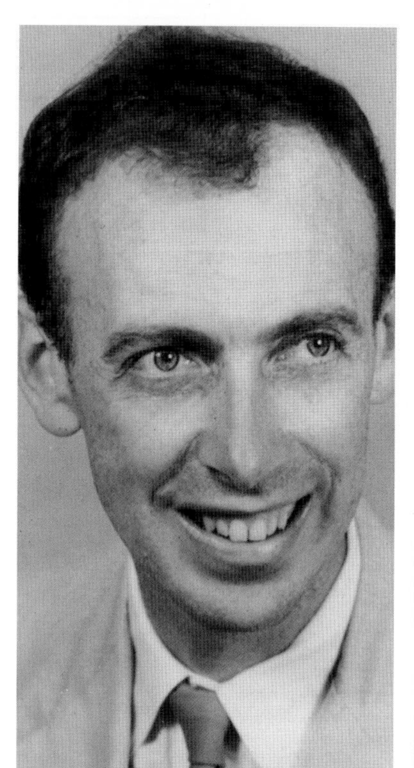

Genetic Breakthrough
James Watson

1953

Genetic Breakthrough

In 1953, American biochemist James Watson and English biophysicist Francis Crick figured out the structure of deoxyribonucleic acid, or DNA. This substance carries the genetic code—instructions for every living cell. In April, Crick and Watson published their discovery in the British scientific journal *Nature*. Their findings marked the birth of modern genetic science and earned them the Nobel Prize for Medicine in 1962.

1953

Piltdown Hoax

The Piltdown Man fossil was unearthed near Piltdown, England, and reported in 1912. This archaeological find thrilled scientists. It also turned the world's understanding of **evolution** upside down. The skull found seemed to fill the gap between the evolution from apes to man. The finding caused

Piltdown Hoax

1956
The Nobel Prize in chemistry is awarded to researchers studying how chemical reactions work.

1957
The first human-made satellite is placed in orbit.

1958
Earth's magnetosphere is discovered.

a great deal of controversy and debate around the world. Then, in 1953, the skull was proven to be a hoax. New methods of dating artifacts led scientists to determine that the so-called Piltdown Man was really made from the jaw bone of an orangutan and the skull of a human. The two were put together, stained to look old, and buried to make it appear that they went together. After testing, scientists found that the bones were from two different ages. Some people blamed Charles Dawson for the fraud. He had found the fossils in the first place. Others said Dr. Arthur Smith Woodward was involved. They said the finding supported his theories. In the end, the truth was never uncovered, and Piltdown Man was disregarded.

Piltdown Hoax

Into the Future

Nuclear power can both help and hurt human beings. Today, the United States has more nuclear weapons and power plants than any other nation. Nuclear plants generate power without greenhouse gases but create toxic by-products. Do you think the United States should use more or less nuclear power?

1959

The Dyson Sphere, a structure large enough to contain a star, is introduced.

1960

TIROS-1, the world's first weather satellite, is launched by the United States.

Science and Technology 1940s

1940

Dangerous Elements

In 1940, U.S. scientists discovered plutonium. This element would later be used in the first nuclear weapon. The weapon that would bring down the **Axis powers** was made possible, ironically, by the work of Germans. Two German scientists had bombarded uranium with neutrons. They later discovered how to split the uranium atom in two. American scientists took the experiment further and found how to produce plutonium. This new element could be split easier and faster than uranium. This meant that when used in weapons, it was very powerful. Just 11 ounces of plutonium could create an explosion equal to 20,000 tons of dynamite.

Dangerous Elements

1941
The element Plutonium is created for the first time in the laboratory.

1942
Scientist Stephen Hawking is born.

1943
Scientist Oswald Avery proves that genetic data is carried by DNA.

1942

Manhattan Project

Manhattan Project

In 1942, the U.S. government decided that people doing research on nuclear weapons should combine their efforts. Physicist J. Robert Oppenheimer and Brigadier General Leslie R. Groves worked together to lead the Manhattan Project, named for the location of its headquarters. About 100,000 people and a dozen university laboratories contributed to the research. Top scientists were then invited to a secret New Mexico compound to help develop nuclear weapons. Oppenheimer handled the scientific side of the project, and Groves was in charge of the staff and defense of the site. Less than three years later, the efforts paid off. A test explosion flashed brilliant light across the desert. Oppenheimer was concerned about the effects of this weapon, seeing it as the "shatterer of worlds." Groves was excited. The project could end the war. Their research would be applied to create the most devastating weapon in history—the atomic bomb.

1944

Tuberculosis Controlled

By 1944, penicillin was helping people fight infections. It was not, however, effective against the disease tuberculosis (TB). In January 1944, microbiologist Selman Waksman announced his discovery of a new antibiotic called streptomycin. Testing confirmed that it worked well against tuberculosis. This discovery changed the treatment of TB forever—doctors could attack the disease rather than the symptoms. Scientists soon discovered that streptomycin was not enough to fight TB on its own. Still, it worked well when combined with other antibiotics. The resulting drug was so effective that, within five years of introducing it, TB rates dropped dramatically in many parts of the world.

Tuberculosis Controlled
Selman Waksman

1944
Aviator Howard Hughes sets a record time flying the Lockheed Constellation across the United States.

1945
The Smyth Report informs citizens of the uses of atomic power.

ENIAC Built

1946

ENIAC Built

In 1946, scientists at the University of Pennsylvania created the world's first all-purpose electronic digital computer. The machine, called the Electronic Numerical Integrator and Computer, or ENIAC, weighed 30 tons and contained 18,000 vacuum tubes. It was built to perform ballistic equations for the U.S. Army. ENIAC was installed at Aberdeen, Maryland. Despite its enormous size, ENIAC was an improvement over previous computers. The machine could solve 5,000 mathematical operations per second. People in the science community were excited by what ENIAC could mean for the future. A magazine predicted that computers in years to come might have only 1,000 tubes and might weigh no more than 1.5 tons.

Yeager Breaks Barrier

1947

Yeager Breaks Barrier

Before 1947, the speed of sound seemed too fast to imagine. When a plane flies slower than the speed of sound, it sends sound waves ahead of it. As the plane nears the speed of sound, these waves can cause pressure and spin the plane out of control. In October, Charles Yeager, one of the country's best fighter pilots, boarded the Bell X-1 plane to attempt to break the sound barrier. This research plane was designed to reduce the amount of pressure on it. The bullet-shaped design was a success. The 24-year-old pilot reached speeds of 700 miles an hour—or Mach 1.06—at 43,000 feet in the air. Despite two broken ribs from an accident the night before, Yeager became the first person to fly faster than the speed of sound.

1946 The U.S. Navy's Blue Angels give their first performance.

1947 The basic principles of quantum mechanics are discussed in New York.

1948 The Mile High Stadium in Denver, Colorado, is completed.

1947

Transistors Revolutionize Electronics

In 1947, William Shockley invited some colleagues at Bell Telephone Laboratories to look at the result of his experiments. Shockley and co-researchers John Bardeen and Walter Brattain showed how an electric current could pass through a tiny device called a transistor. Their discovery put an end to Bell's glass-enclosed vacuum tubes, which had been used until that time. Transistors were inexpensive to produce, long-lasting, did not require much power, and did not create as much heat as vacuum tubes did. When the discovery was made public, electronics experts were excited by the possibilities. Within a decade, transistor radios that could fit in a pocket demonstrated how far the invention had come. Brattain, Shockley, and Bardeen were awarded the Nobel Prize for Physics in 1956.

Transistors Revolutionize Electronics
John Bardeen, William Shockley, and Walter Brattain

Into the Future

Transistors were one of the earliest inventions that allowed electrical devices to be made smaller. Many people today consider the small size of a device, such as a cell phone or music player, an advantage. Think about devices you use that could be improved by being made smaller. Are there any that would not work as well if they were too small?

1949

Radiocarbon dating, a method used to determine the age of an object, is discovered.

1950

The first U.S. Air Force jet fighters see combat in the Korean War.

Science and Technology 1930s

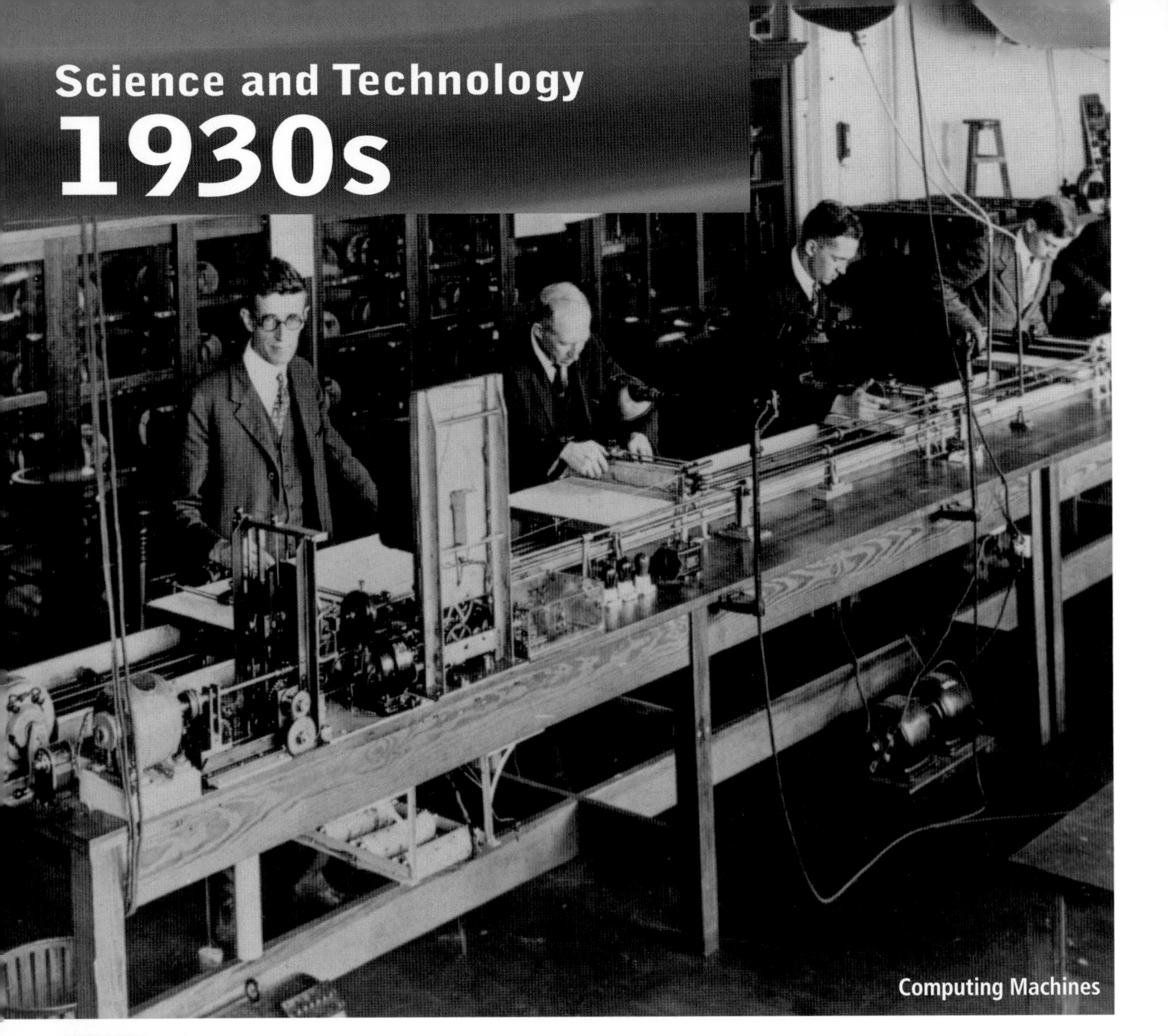

Computing Machines

1930

Computing Machines

Engineer Vannevar Bush and a team of scientists at Massachusetts Institute of Technology in Boston started working on a new machine in 1930. Their "differential analyzer" was an improvement over mechanical adding machines and was the father of electronic calculators. It was the first analog computer and took up hundreds of square feet of floor space. Hundreds of steel rods rotated to solve calculations, and programmers used hammers and screwdrivers, not keyboards, to make adjustments for different problems. The computer could solve complicated problems with up to eighteen independent variables at once. Bush's invention was improved on during World War II. Scientists abandoned his electro-mechanical methods in favor of strictly electronic technology. Although Bush's differential analyzer was a long way from today's computers, it helped lay the foundation for them.

1931

Inventor Thomas Edison dies.

1932

Experimental evidence is found to support Einstein's Theory of Relativity.

1931

Amazing Aviators

Pilot Wiley Post made his dream come true in 1931. He flew around the world in less than nine days. He borrowed his boss's plane and made changes to the engine so it could fly longer. He and navigator Harold Gatty flew from New York to Newfoundland on June 23 and then flew to Liverpool, England. From there, they met thousands of cheering fans in Berlin, Germany, flew through a storm in the Soviet Union, and landed in Alaska. They crossed the Rocky Mountains in Canada and landed on a street serving as a runway in Edmonton, Alberta. Then, the duo returned to New York only eight days, fifteen hours, and fifty-one minutes after taking off. About 10,000 excited onlookers cheered as the plane touched down.

Another 1930s aviator was not as lucky. Amelia Earhart's achievements included being the first female pilot to fly across the Atlantic Ocean alone in 1928 and the first woman to receive the Distinguished Flying Cross in 1932. On June 1, 1937, she and navigator Frederick Noonan set out to become the first to fly around the world along the equator. The plane touched down in Puerto Rico, flew over the northern coast of South America, crossed Africa and India, and covered Southeast Asia. On June 28, Earhart and Noonan reached Australia. On July 2, the two started their trek over the ocean to Howland Island. They never made it. The plane disappeared without a trace. Explanations ranged from the plane having run out of gas and crashing to Earhart having been shot down while on a secret spy mission. The disappearance remains unsolved and a mysterious part of thirties history.

Amazing Aviators
Amelia Earhart and A.N. White.

1933
The concept of a neutron star is suggested.

1934
A special film canister is made for 35mm film.

1935
The Richter Scale is made for calculating the intensity of earthquakes.

1938

Hungarians to the Point

Since the 1800s, people had been looking for another way to write. They were hoping to find a replacement for pen and ink. In 1938, their hopes were answered when the Hungarian brothers Ladislao and Georg Biro invented the ballpoint pen. This new pen had a narrow tube of ink emptying into a steel ball at its end. The design allowed ink to spread evenly from the ball to the writing surface. The Biro pens were advertised as the only pens that could write underwater—a gimmick that was tested and proved outside countless department stores. An Austrian chemist named Fran Seech later improved the design so that the ink would dry as soon as it touched the paper. The Biro company was taken over by the French company Bic, which became well known for its throw-away ballpoint pens.

"Evolutionary" Ideas

1937

"Evolutionary" Ideas

Geneticist Theodosius Dobzhansky brought Gregor Mendel's ideas of genetics together with Charles Darwin's theories of evolution. Before Dobzhansky's research, many scientists thought that evolution and Darwin's natural selection took place very slowly over a long period of time, producing a single, ideally adapted species. Dobzhansky showed that there was a great deal of difference between members of the same species in the same environment. He used vinegar flies to test his ideas. Some members of the species had genes that were apparently of no use to them in their environment. The flies were genetically quite varied, which

Hungarians to the Point

1936
The first television and telephone cables are laid between New York and Philadelphia.

1937
The Golden Gate Bridge in San Francisco is opened to the public.

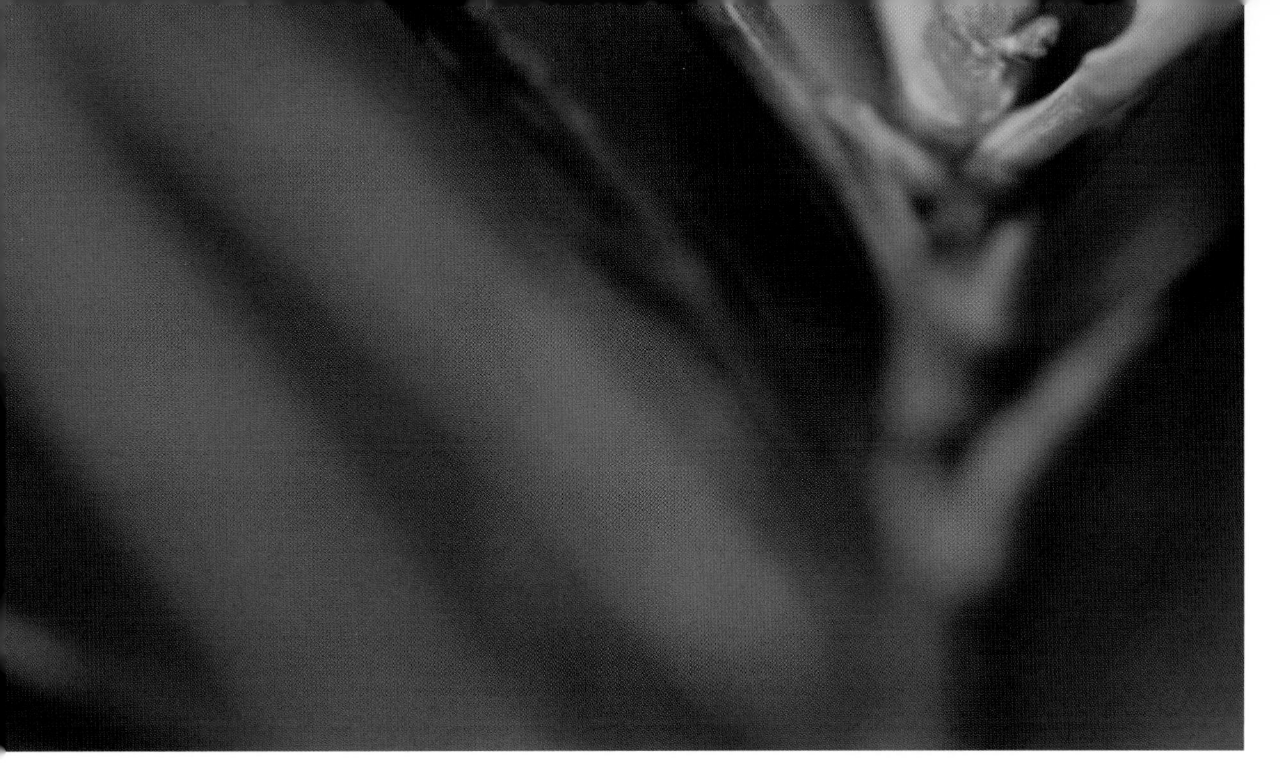

would make the species better at adapting to changes in its environment. This variation suggested that some members of the species would be better equipped to survive a major environmental change. Dobzhansky published his findings in his 1937 book entitled *Genetics and the Origin of Species*.

Into the Future

Knowing that living things adapt has helped doctors develop new medicines to treat diseases that change over time. For example, doctors need to create a new flu vaccine every year because the virus changes often. What other ways can doctors use this knowledge to address illness and disease?

1938

A meteorite weighing more 400 tons falls to Earth near Chicora, Pennsylvania.

1939

Nuclear fission is discovered.

1940

FM radio waves are demonstrated to the U.S. government.

Science and Technology
1920s

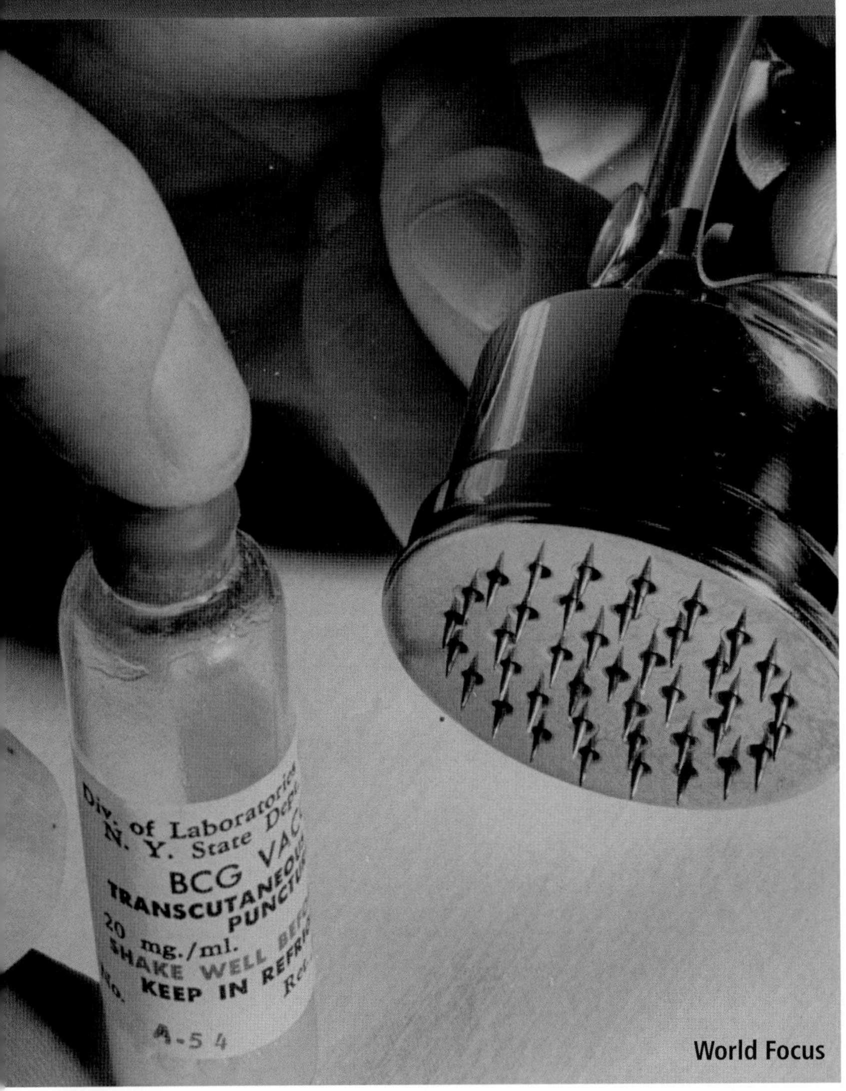

World Focus

1921

The Fight Against Tuberculosis Begins

One of the most deadly diseases in the world in the twenties was tuberculosis. In 1921, after fifteen years of experimenting, a vaccine was found to increase children's resistance to the horrible disease. French scientists Albert Léon Calmette and Camille Guérin created the vaccine from a weakened strain of live tuberculosis bacteria. They injected schoolchildren with this strain to stop the spread of this incurable disease. Most European countries accepted the vaccine, but the treatment was not offered in the U.S. and some European countries. American scientists were nervous about the vaccine—in Germany in 1930, 73 of the 249 infants vaccinated died within a year of being injected. The U.S. and British authorities accepted the vaccine in 1940. By the time of Guérin's death in 1961, 200 million people worldwide had taken advantage of the tuberculosis vaccine.

1922

Hollywood in Technicolor

In 1922, black and white movies were given a splash of color. Herbert T. Kalmus and Daniel F. Comstock had begun to tackle the challenge of colorizing film in 1915 with their company, Technicolor Motion Picture Company. They soon thought of using a camera that had two strips of film—one for green and the other for red—and a prism to separate the light into the primary colors. The strips were brought together when the film was developed and printed. This was exciting because no special **projectors** or filters were needed to achieve color. The first movie made using the Technicolor process was the 1922 remake of Madama Butterfly called *Toll of the Sea*. The film met great

1921
Insulin is discovered.

1922
Henry Berliner shows an early helicopter to U.S. military officials.

reviews, but the new color did not catch on. It was too expensive for producers to use. Technicolor was often used to bring color to special sequences or scenes in otherwise black and white films until around 1930.

Three-color Technicolor was developed and first used by Walt Disney in his animated films. The full-length movie *Becky Sharp* (1935) was the first film to use the three-color process.

Hollywood in Technicolor

1923

Icy Invention

Americans were used to having the iceman track mud through their kitchens as he delivered blocks of ice for their iceboxes. Experiments with mechanical refrigeration started in the early 1900s. Businesspeople, including fur-vault owners and dairy operators, were looking for a better way to keep their products cool. This led to advances in home refrigeration, too. In 1923, Frigidaire, a division of the huge company General Motors, revolutionized U.S. kitchens. Its new machine had a separate cabinet for the icebox, complete with machinery to keep it cool. The refrigerator was small, neat, and convenient. Soon, nearly every family scrambled to buy a Frigidaire.

By 1944, about 85 percent of American families owned one. This invention put an end to the muddy footprints and to iceboxes forever.

Icy Invention

1923
Chuck Yeager is born.

1924
Edwin Hubble announces the existence of other galaxies.

1925
Air conditioning is first used in movie theaters.

1927

Iron Lung Helps Americans Breathe

Poliomyelitis, or polio, was a terrible disease that struck the central nervous system. Young children were at risk of contracting this disease. Before 1927, doctors could do little for patients as polio paralyzed their lungs. The patients slowly suffocated. In 1927, Harvard physician Philip Drinker created a machine that helped people breathe. This artificial breather was nicknamed "the iron lung." It moved air in and out of a chamber with the help of a vacuum pump. Patients would lie within the tank-like chamber, and the movement of the air forced the lungs to inflate. Patients were kept alive in these iron lungs until they recovered from polio. This invention saved the lives of many polio victims until a vaccine in the fifties essentially wiped out the disease in North America.

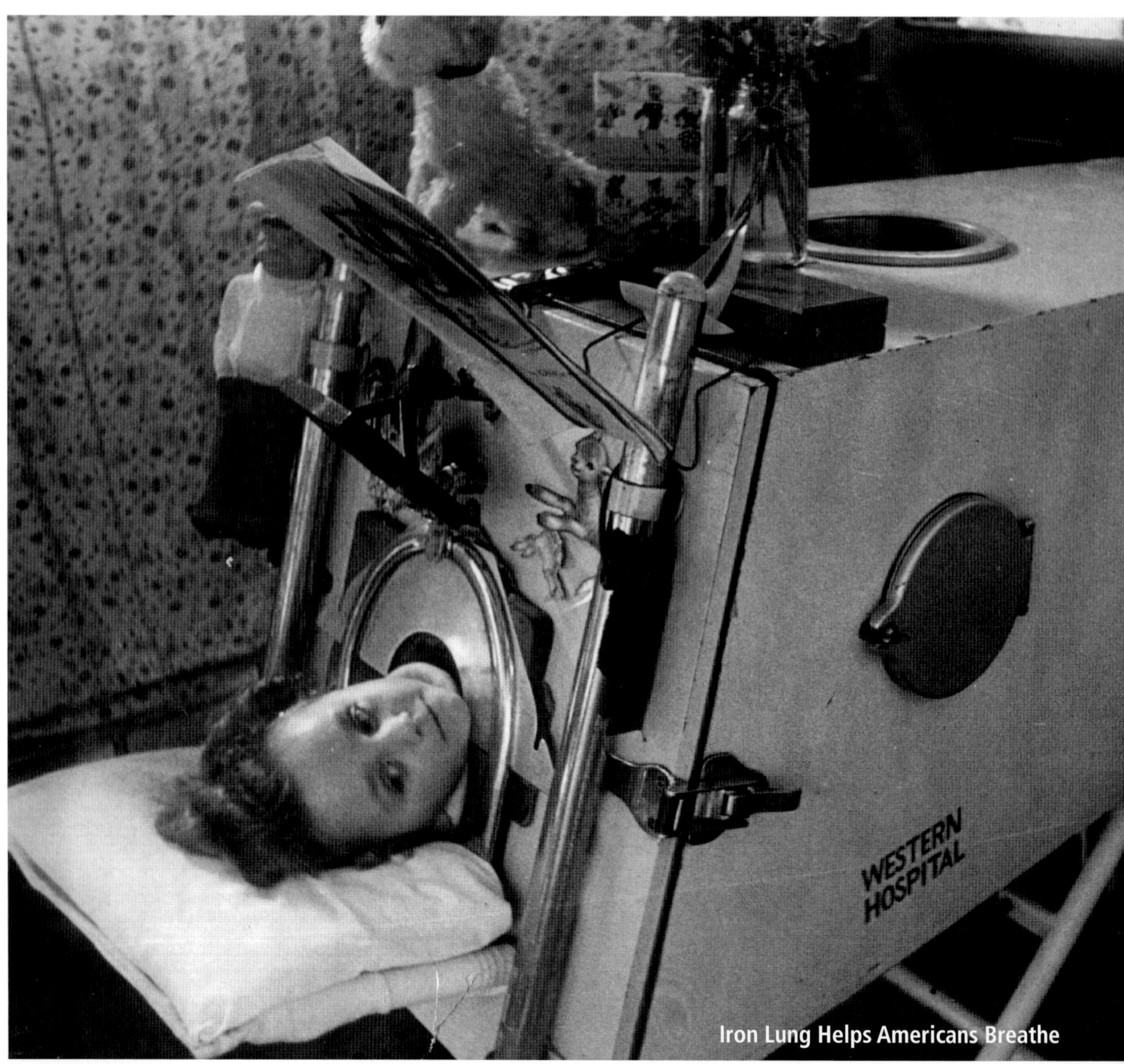

Iron Lung Helps Americans Breathe

1926

The first liquid-fueled rocket is launched in Auburn, Massachusetts.

1927

The aircraft carrier USS *Lexington* is commissioned.

1927

Aviator Crosses the Atlantic

Aviator Crosses the Atlantic

Other people had flown across the Atlantic Ocean, but Charles Lindbergh was the first to do it alone. A French hotel owner in New York was offering a $25,000 prize to anyone who flew nonstop from New York to France, and Lindbergh decided to give it a try. On May 20, 1927, he flew out of Long Island, New York, at 7:54 am. He landed at Le Bourget airport in Paris thirty-three and a half hours later.

He had not slept the night before the flight, so he had to fight to stay awake during the 3,614-mile journey. Fog, storms, and instability due to the light weight of the plane added to his troubles, but he touched down safely. About 100,000 cheering fans were there to greet him. Many of them wanted a souvenir of the incredible journey, and the *Spirit of St. Louis* plane was nearly torn apart. In June, when Lindbergh returned to the U.S., four million people were gathered to welcome him. Lindbergh used his fame to promote commercial overseas air travel. By 1935, Pan Am began offering passengers flights across the Pacific Ocean. Lindbergh's dream was a reality.

Into the Future

Without the efforts of aviators, it would be much more difficult to travel around the world. Without the aid of airplanes, business, vacations, and even sporting events would all have to take place much closer to home. Think about other ways humans can travel from place to place. What do you think will be the next big transportation invention?

1928
The first successful public television broadcast is made in New York.

1929
Nuclear reactions are first suggested as the energy source of stars.

1930
Pluto is discovered.

Science and Technology
1910s

Halley's Streaks across the Sky

1910

Halley's Streaks across the Sky

On May 18, 1910, more people were on their rooftops than inside their homes. Many people were afraid of what they saw. Others were fascinated. The object of these mixed emotions was Halley's Comet. Astronomers and scientists had been eagerly awaiting the arrival of the comet for decades. It appeared approximately every seventy-six years. A German astronomer finally located the comet in 1909, and people waited for it to become visible to the naked eye. Many citizens were not as eager. They did not understand the comet and believed that gases from the comet's tail could cause serious damage to Earth's atmosphere and could even harm people. Tabloid newspapers fed the fear around the world with terrifying and inaccurate stories about the comet. Some people refused to go to work around the expected time of the comet's arrival—they wanted to stay with their families. In reality, the trail of gases stayed about 250,000 miles from the surface of our planet. Astronomers collected any information they could from the passing comet, knowing that

1911

An aircraft lands on the deck of a ship for the first time in San Francisco Bay.

1914

The world's fifth-largest refracting telescope is built in Pittsburgh, Pennsylvania.

Start Your Engines

they would have to wait many years to study it again.

1911

Start Your Engines

Before 1911, automobile drivers had to use hand cranks to start their vehicles. Charles Franklin Kettering's electric starter put an end to that unpleasant, unreliable, and potentially dangerous practice. Many people thought that such a device was impossible. They figured that a motor capable of turning over a car engine would weigh too much and would not allow the car to carry passengers. Kettering recognized this problem as well, but he also realized that the starting motor did not have to run the automobile. He created a small generator and motor unit. The motor was run by a storage battery. It had just enough power to start the engine. Once the engine was running, it powered the generator, which powered the battery. Kettering had proven the world wrong. His system was a success, and Cadillac became the first company to use self-starting cars.

1910

Diagnosis Affects Thousands

African Americans had been suffering from a blood condition that no one could identify. The condition led to serious infections, injury to major organs, and extreme pain throughout the body. In 1910, a Chicago physician figured out the problem. Dr. James Bryan Herrick examined a West Indian student who had these symptoms. He found that the student had unusually shaped red blood cells. These cells looked like a crescent-shaped farm tool called a sickle. The cells could not pass through tiny blood vessels, so oxygen and other nutrients were withheld from organs and other parts of the body. This is what caused the pain and organ damage. Herrick called the disorder sickle-cell anemia. This disease was in about 3 percent of the African-American population.

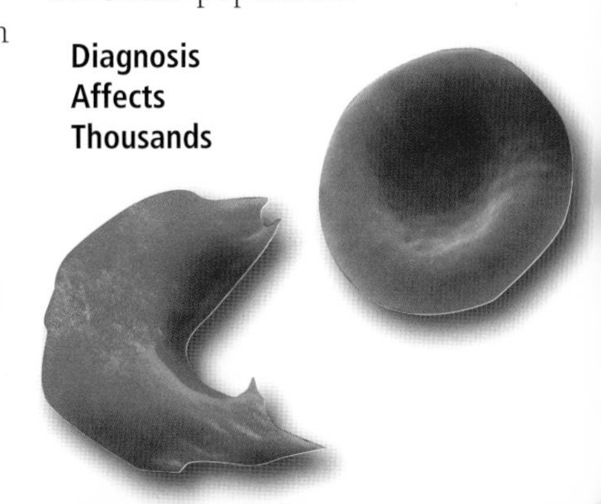
Diagnosis Affects Thousands

1915
The Pangaea theory, that at one time all of Earth's continents were joined together, is proposed.

1918
Charles Strite invents the pop-up toaster.

1919
Airplanes make nonstop crossings of the Atlantic.

Science and Technology 1900s

1900

Army Experiments

Dr. James Carroll became very ill on August 27, 1900. His colleague Dr. Jess William Lazear became ill as well. The two were part of a team headed by Major Walter Reed sent by the U.S. army to study mosquitoes in Cuba. They set out to prove that yellow fever was the result of mosquito bites. Volunteers from the ranks stepped forward to take part in the experiments. Private John R. Kissinger and others like him refused to accept the $250 payment for the experiments—they felt it was their duty, and they were acting in the interests of humankind. Kissinger was paralyzed, Carroll recovered with heart damage, and Lazear died. Their sacrifices saved the lives of millions. Through the experiments, Reed proved that the mosquito species *Stegomyia fasciata* carried the disease. These insects picked up yellow fever from victims by drawing blood in the first three days of the individual's sickness. Then it took another twelve days for their bite to become potent. Yellow fever can lead to delirium, coma, or even death.

1904

Bigger Is Better

The world's first electric underground railway was built in London, England, in 1890. It was a celebrated feat, but New York City topped it. It built the world's biggest subway in October 1904. This system of transportation was a necessity. Aboveground traffic had become too crowded and would only get worse as the population grew. The public rail system was a popular solution to the problem. The train ran from the Brooklyn Bridge uptown to Broadway at 145th Street. On its first day, about 500,000 people stepped aboard to experience the underground system for themselves. The cars sped along the 15 miles of track at 25 miles per hour. As promised, the subway system took New Yorkers from Harlem to City Hall in fifteen minutes. It was not a complete success, though. The mass of people created an opportunity for crime. On the very first day, at 7:00 in the evening, a thief stole a $500 diamond pin from a subway passenger. Despite the risks, the subway was a welcome escape from the heavy traffic on the streets of New York.

Army Experiments

Bigger is Better

1900

The first successful airship is constructed.

1904

Vacuum tubes, used in many early televisions and computers, are invented.

1908

Model T

Although thousands of Oldsmobiles were motoring along U.S. streets by 1902, these were luxury cars, and most people could not afford them. Then, in 1903, automobile engineer Henry Ford took a chance. He formed the Ford Motor Company. Ford did not want his automobiles to just be toys for the rich, so he began to make his own parts rather than simply assembling parts from other companies. This cut his costs and allowed him to make less expensive cars. By 1906, Ford's cars were being made from steel, making them stronger, lighter, and faster than any other. For a year, Ford and his team worked on a new design. The result was the Model T, introduced in October 1908. For $850, Americans could buy a sturdy vehicle that stood up to potholes. Ford sold 18,664 cars from 1909 to 1910. The number nearly doubled the following year. In the 1910s, Ford refined his methods, creating the automobile assembly line. This meant he could offer cars for even less money.

1908

Flying High

Orville and Wilbur Wright's bicycle shop developed into something else altogether—an air travel laboratory. The Wrights decided that they would design an airplane and learn to fly it. Over the next few years, the brothers studied **airborne** objects. In 1900, they built a kite that could support a pilot. They tested their inventions and continued to try out new models. On December 17, 1903, the Wright brothers thought they had eliminated all potential problems. They were right. The Wright biplane took flight at Kill Devil Hills near Kitty Hawk, North Carolina. This new design had a wingspan of 39 feet and weighed 750 pounds, pilot included. Orville made the first successful flight, staying in the air for twelve seconds.

In 1904, the Wrights added a 16-horsepower engine to their plane. They were able to fly farther and execute tighter turns in this machine. By 1905, the brothers been contracted to design airplanes for the U.S. War Department. In the following years, they patented their invention and looked for buyers. In 1908, they gave the first public showing of the planes in the U.S. Orville set a record by keeping the plane in the air for more than an hour on September 9. They then showed their planes in Europe, and many countries took notice. With orders flowing in, the brothers established the airplane-manufacturing Wright Company.

Model T

Flying High

1905

***Tyrannosaurus rex* is first named and described.**

1906

The first radio broadcast is made.

1910

The exact location of genes in cells is discovered.

ACTIVITY

Into the future

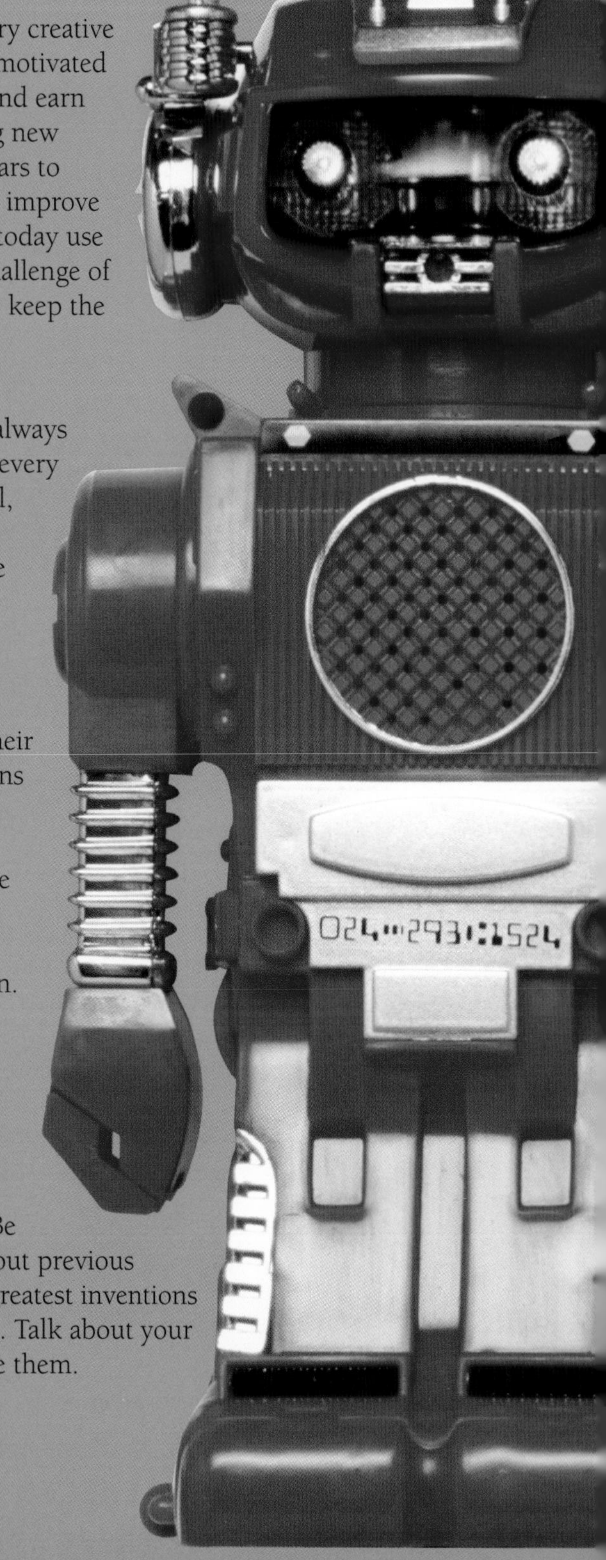

Many technological advances have come from very creative people. Inventors and entrepreneurs are often motivated by desires to both improve the lives of others and earn wealth for themselves. Some accomplish this by creating new ways of doing old tasks, such as making airplanes and cars to transport people faster than ever before. Others work to improve previous inventions, just as communication companies today use technology in the space race. Inventors often face the challenge of designing technology that is energy-**efficient** in order to keep the planet healthy for future generations.

We live in an era of advanced technology, but there are always better ways of doing things. Think about things you do every day. Is there a better way to wash dishes, travel to school, or style your hair? If you can think of a better way, and the technology that would let you do it that way, you are on the right path to coming up with a new invention.

Some inventions do not build on earlier ideas. The Wright brothers had no previous models to experiment with when designing the first airplane. They had only their observations of flight in nature and the belief that humans could achieve it as well. Think about something that humans are still unable to do with modern technology. What kinds of things might make our lives better but are beyond our reach? How might technology be used to accomplish these things? Answering these questions is the first step toward coming up with your own invention.

Become an Inventor

Once you have answered these questions, start thinking about ways to use technology. Write about or draw a diagram of an invention that will help make life easier. Be specific about how it will work. Include information about previous inventions, if any, that inspired your idea. Many of the greatest inventions in history were developed by more than just one person. Talk about your inventions with your friends, and share ideas to improve them.

FURTHER Research

Many books and websites provide information on science and technology. To learn more about this topic, borrow books from the library, or surf the Internet.

Books

Most libraries have computers that connect to a database for researching information. If you input a key word, you will be provided with a list of books in the library that contain information on that topic. Non-fiction books are arranged numerically, using their call number. Fiction books are organized alphabetically by the author's last name.

Websites

To learn more about science, visit **www.sciencenewsforkids.org**.

For science experiments, surf to **http://pbskids.org/zoom/activities/sci**.

Glossary

airborne: carried in the air

astronomer: a person who studies space and the universe beyond Earth

atmosphere: the gases that surround the planets

atom: the smallest unit of a chemical element

Axis powers: Germany, Japan, Italy, and their allies during World War II

efficient: to perform well with little wasted effort

elliptical: having an oval shape

enzyme: a substance that helps chemical reactions occur inside a living thing

evolution: a theory about how all living things develop and change throughout history

genes: chromosomes, or units of heredity, that are given from parent to child

gyroscope: a device that is mounted in one place and has a wheel that can be turned in any direction

immune system: a series of organs and cells that protect the body from illness and disease

microprocessor: a circuit that contains all the functions of a certain type of computer

nuclear energy: energy that comes from the nucleus, or center, of an atom

programmers: people who write computer programs

projectors: machines used to show images on a screen by shining light through film

self-contained: having everything one needs to live or function

thermonuclear: having to do with nuclear reactions that take place at very high temperatures

vaccines: germs that are not active and are used to protect a person from acquiring a specific illness

Index